Risk Assessment Using FMEA

A Case of Reliable Improvement

Copyright ©2021 Mohammed Soliman

All rights reserved

# By Mohammed Hamed Ahmed Soliman

While every precaution has been taken in the preparation of this book, the publisher assumes no responsibility for errors or omissions, or for damages resulting from the use of the information contained herein.

RISK ASSESSMENT USING FMEA: A CASE OF RELIABLE IMPROVEMENT

First edition. April 1, 2021.

Copyright © 2021 Mohammed Hamed Ahmed Soliman.

Written by Mohammed Hamed Ahmed Soliman.

CONTENTS

Introduction to Risk Management 6

Asset Criticality Assessment 10

FMEA as a Tool to Minimize Risks and Improve Reliability 16

Failure Mode 21

FMEA Team 25

Steps of FMEA Process 27

Failure Modes, Effects, and Causes 44

A Case of Reliable Improvement 58

   Improving the Reliability of a Pump Station System 58

FMEA and Continuous Improvement 78

Appendix.1 Brainstorming 81

Appendix.2: Cause and Effect Diagram 95

References 99

About the Author 101

INTRODUCTION TO RISK MANAGEMENT

What is risk management?
Is the way toward recognizing, evaluating and controlling dangers to an association's capital and profit. These dangers, or threats, could come from a wide assortment of sources, including monetary vulnerability, legitimate liabilities, vital administration mistakes, mishaps and catastrophic events.

Determination of risk level
The probability is the likelihood of an event occurring and the consequences, to which extent the project is affected by an event, are the impacts of risk. By combining the probability and impact, the Level of Risk can be determined.

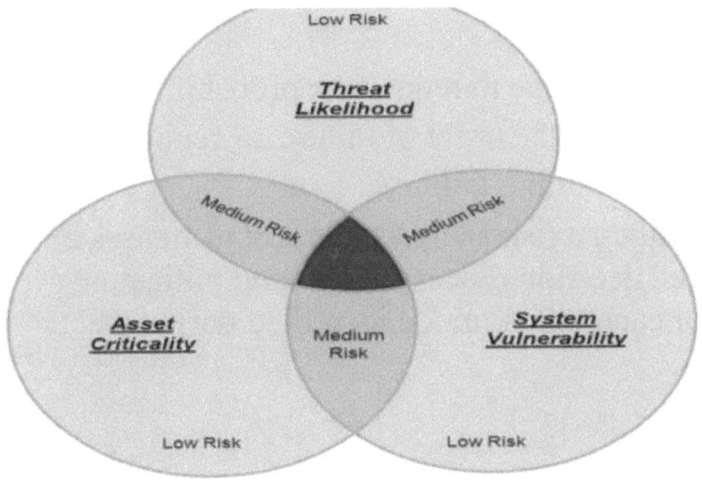

	Very Likely		Average		Very Low
No Impact	Priority 5	Priority 5	Priority 5	Priority 5	Priority 5
	Priority 3	Priority 3	Priority 3	Priority 5	Priority 5
Some Impact	Priority 2	Priority 2	Priority 2	Priority 5	Priority 5
	Priority 1	Priority 1	Priority 2	Priority 4	Priority 5
Disastrous Impact	Priority 1	Priority 1	Priority 2	Priority 4	Priority 5

You can always design this matrix for your system components to determine the risk priority value for each component.

Knowing which component of the system is likelihood to fail is very useful in determining which strategy to use to prevent this failure. But first, you need to identify your priorities, which system component is likelihood to fail and need attention first, and if this component is critical to your business, your process or not.

Here are some few questions that help define the criticality of the system components:

The RCM analysis carefully considers the following questions:

• What does the system or equipment do; what is its function?
• What functional failures are likely to occur?
• What are the likely consequences of these functional failures?
• What can be done to reduce the probability of the failure, identify the onset of failure, or reduce the consequences of the failure?

In performing a risk analysis, the risk team uses a structured decision process to develop mitigating tasks for each failure mode identified during the analysis:

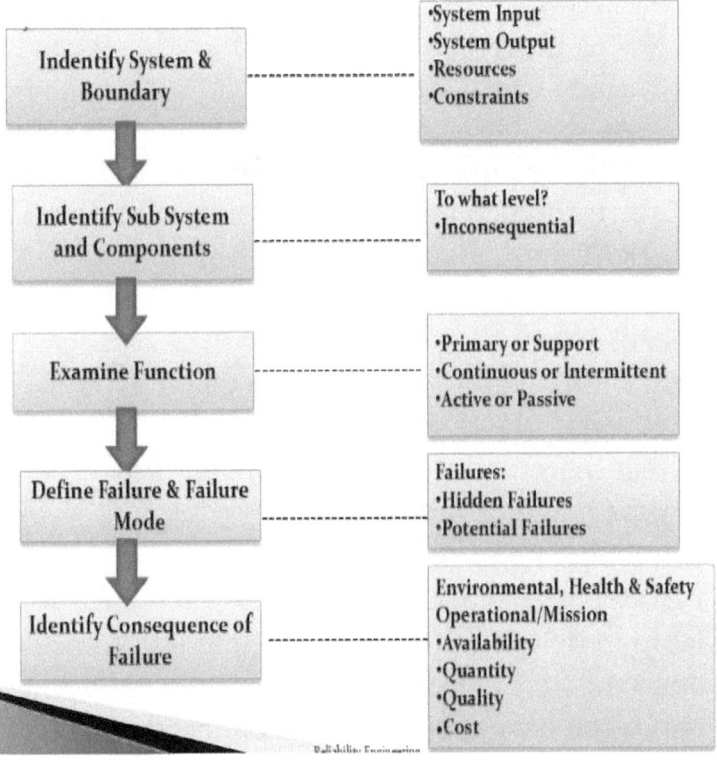

ASSET CRITICALITY ASSESSMENT

Importance of Equipment Criticality Analysis:
1. Influence the priority assignment of the Work Orders.
2. Influence the Work Orders execution speed.
3. Determine which Maintenance Class should come first.
4. Effect the scheduling of the preventive maintenance program.
5. Influence the priority of the preventive maintenance work.
6. Help in determining which maintenance approach to be used (corrective, preventive, condition based, proactive, risk based...etc.).

Criticality Measuring Principles

1. Safety
- ✓ Safety equipment (equipment carrying peoples, firefighting system, and alike).
- ✓ Consequence of parts failure effect safety (elevator ropes broken, firefighting generator stopped...etc.).
- ✓ Gas piping leakage at any point poses a risk.

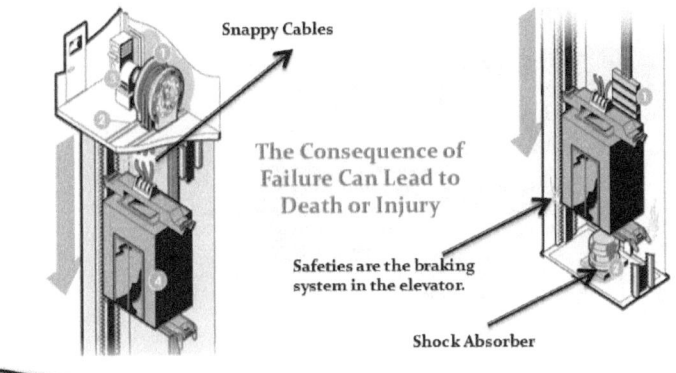

2. Production, Process
✓ Equipment breakdown affect the whole production line.
✓ Equipment breakdown affect partially the production line.
✓ Equipment failure has native effect on production quality.

The Consequence of Failure Can Lead to Process Shutdown, Major Losses, or Quality Problem.

Criticality decreases with redundancy in the system.

Criticality is influenced by the availability of standby equipment in a system but how much time does it take for the standby equipment to operate? And what is the effect of this on the production?

3. Time (Time=Money)
1 min production=How much?
1 Hr. production =How much?

If your equipment is classified as critical, ask yourself the following questions:
- What is the preventive maintenance program I have for it? Enough? Or not?
- How much time does it take to repair it in case of failure?
- Spare parts allocation? Available? Not available? If not available, how much does it take to allocated it from the vendor?

- Require special skills for repair? Is my team trained to repair it in a proper time?
- Do you have an emergency plan for it in case of accident or failure? Is your team aware of this plan and trained on it?

4. Money, Cost

Equipment that fails in service can cost up to 10 times more to repair than the equipment repaired when predicted by condition monitoring.

Failure Consequence Criticality Classification

Class	Health/ Safety/ Environment	Production	Cost
High	•Potential for serious personnel injuries •Render safety critical system inoperable •Potential to fire in classified area •Potential for large pollution	Stop in production/significant reduce rate of production exceeding X hours	Substantial cost
Medium	•Potential for injuries requiring medical treatment •Limited effect on safety system •No potential for fire in classified areas •Potential for moderate pollution	Brief stop in production/reduced rate of production lasting less than X hours	Moderate cost
Low	•No potential for injuries •No potential for fire or effect on safety system •No potential for pollution (specify limit)	No effect on production within a defined period of time	Insignificant cost

Risk Class, SAFETY first!

Level	Description	Example
A	Major effect on HSE	Fire fighting generator
B	Major effect on process=high downtime cost	Furnace generator
C	Normal effect on HSE Normal effect on process Without Standby	
D	Normal effect on process Normal effect on HSE With Standby	

FMEA AS A TOOL TO MINIMIZE RISKS AND IMPROVE RELIABILITY

What is FMEA?

Failure mode and effect analysis (FMEA) was initiated by the aerospace industry in the 1960s to improve the reliability of systems. It is a part of total quality management programs and should be used to prevent potential failures that could affect safety, production, cost or customer satisfaction. FMEA can be used during the design, service or manufacturing processes to minimize the risk of failure, improving the customer's confidence while also reducing costs.

A FMEA is a systematic method for identifying and preventing product and process problems before they occur. FMEAs are focused on preventing defects, enhancing safety and increasing customer satisfaction.

FMEAs are conducted in the product design or process development stages, although conducting an FMEA on existing products and processes can also yield substantial benefits.

What is the purpose of a FMEA?
Preventing the process and product problems before they occur is the purpose of Failure Mode Effect Analysis. Used in both the design and manufacturing process, they substantially reduce costs by identifying product and process improvement early in the develop process when changes are relativity easy and inexpensive to make.

FMEA can provide the answer to many problems:
How can we prevent this problem from occurring again in the future?
How can we minimize the risk of this potential failure?
How can we produce an error-free product?
How can we reduce the warranty costs?
How can we improve the safety condition in the workplace?

FMEA as a part of a Comprehensive Quality System
Can FMEA be used alone? While FMEAs can be effectively used alone, a company won't get maximum benefit without systems to support conducting FMEAs.

Two things are necessary needed:
1. A reliable product or process data. Without this data, FMEA becomes a guessing game based

on opinions rather than actual facts. Without data the team may focus on the wrong failure modes or missing significant opportunities to improve the failure modes that are the biggest problems.
2. Documentation of procedures. In the absence of documents and procedures, people working in the process could be introducing significant variation in to it by operating it slightly different each time the process is run.

> FMEA is one of the ISO 9001:2000 requirements as you must have a system capable of controlling process that determine the acceptability of your

Benefits of Failure Modes Effect Analysis "FMEA"

The object of an FMEA is to look for all of the ways a process or product can fail. A product failure occurs when the product does not function as it should or when it malfunctions in some way.

Contribute to improve design for product & process
- Higher reliability.
- Better Quality.
- Increase Safety.

Contribute to cost saving
- Decrease development time & redesign cost.
- Decrease warranty costs.
- Decrease wastes.

Contribute to continuous improvement

FMEA Applies to: System, Process, Design, and Service

FMEA helps manufacturing engineers control the process and eliminate errors during production, thus decreasing warranty costs and wastes.

Service engineers use FMEA to improve the lifecycle of the product and lower its service costs by developing a proper maintenance program.

Potential Applications:
- Equipment components & parts.
- Component proving process.
- Outsourcing/resourcing of product.
- Develop suppliers to achieve quality.
- Major process/ Equipment / Technology Changes.
- Cost Reductions.
- New Product/ Design Analysis.
- Assist in analysis in a flat Pareto chart.

FAILURE MODE

What is Failure Mode?
- Any event which causes a functional failure.
- Ways in which product or process can fail are called failure modes. The FMEA is a way to identify the failures, effects, and risks within a process or product, and then eliminate or reduce them.

Example failure modes:
- Bearing Seized.
- Motor burned out.
- Coupling broken.
- Impeller jammed.

Compressors Failure Modes
Discharge pressure low:
- Air leakage.
- leaking valves.
- Defect gauge.

Engines Failures Mode
Knocking:
 - Pistons hitting the head.
 - Crankshaft plays.
 - Oil pump not function.

Example failure modes, coffee maker

Indeed, even the plain devices have numerous chances for failures. For instance, a trickle espresso producer. A relativity basic family unit machine could have a few things bomb that would deliver the coffeemaker inoperable. Here are a few different ways the espresso make can fizzle:

- The warming component doesn't warm water to adequate temperature to mix espresso.
- The siphon doesn't siphon water into the channel container.
- The espresso producer doesn't turn on consequently by the clock.
- The clock quits working or running excessively quick or excessively moderate.
- There is a short in the electrical rope.
- There is either insufficient or an excessive amount of espresso utilized.

 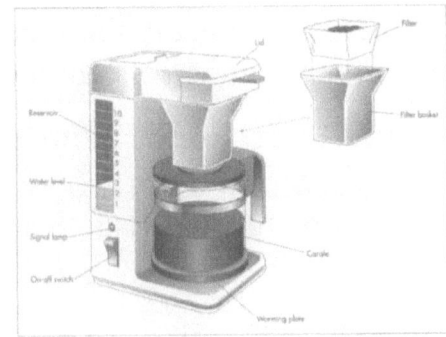

The goal is 100% Customer Satisfaction

Failures are not limited to problems with the product. Because failures also can occur when the user makes a mistake. Those types of failures should be included in the FMEA. Anything can be done to ensure the product works correctly, regardless of how the user operates it, will move the product closer to 100 percent total customer satisfaction. The use of mistake-proofing techniques, also known by its

Japanese term poka-yoke, can be a good tool for preventing failures related to user mistakes.

Failure Effects Description

Local Effect
The failure effect as it applies to the item under analysis.

Ex. Water pump stop.

Next Higher Effect
The failure effect as it applies at the next higher indenture level.

Ex. Water system pressure drop down.

End-Effect
The failure effect at the highest indenture level or total system.

Ex. System stop.

FMEA TEAM

FMEA is not a job for one individual. The best possible results come when teams are composed of contributors from different engineering perspectives. The team should have between four to six members. Team size is determined by the number of areas affected by the FMEA, for example manufacturing, maintenance, design, engineering, material, technical service, etc. The customer adds another unique perspective and should be considered for team membership. If customers cannot be included, the team should devise ways to generate voice-of-the customer data.

Team Leader:
The team leader is responsible for coordinating the FMEA process as follow:
1. Setting up and facilitate meeting.
2. Ensure that all resources are available.
3. Make sure the team moves toward completing the FMEA process.
4. The team leader role is more like of a facilitator rather than decision maker.
5. Determine the boundaries of freedom.
6. Define the scope of the project.

FMEA Team Start-Up Worksheet

FMEA Number:		Date Started:	
Team Members:		Date Completed:	
Leader:			
Who will take minutes and maintain records?			

1. What is the scope of the FMEA? Include a clear definition of the process (PFMEA) or product (DFMEA) to be studied. (Attach the Scope Worksheet.)

2. Are all affected areas represented? (circle one)

YES	NO	Action:	

3. Are different levels and types of knowledge represented on the team? (circle one)

YES	NO	Action:	

4. Are customer or suppliers involved? (circle one)

YES	NO	Action:	

Boundaries of Freedom

5. What aspect of the FMEA is the team responsible for? (circle one)

FMEA Analysis	Recommendations for Improvement	Implementation of Improvements

6. What is the budget for the FMEA?	
7. Does the project have a deadline?	
8. Do team members have specific time constraints?	
9. What is the procedure if the team needs to expand beyond these boundaries?	
10. How should the FMEA be communicated to others?	

STEPS OF FMEA PROCESS

There are several steps to follow when conducting FMEA process:

1. Select a high-risk process, then follow these steps.
2. Review the process: this step usually involves a carefully selected team that includes people with various job responsibilities and levels of experiences. The purpose of an FMEA team is to bring a variety of perspectives and experiences to the project.
3. Breakdown the system into components and sub-components.
4. Brainstorm potential failure modes.
5. List potential effects of each failure mode.
6. Assign a severity ranking for each effect.
7. Assign an occurrence ranking for each failure mode.
8. Assign a detection ranking for each failure mode.
9. Calculate the risk priority number (RPN) for each effect.

10. Prioritize the failure modes for action using RPN.
11. Take action to eliminate or reduce the high-risk failure modes.
12. Calculate the resulting RPN as the failure modes are reduced or eliminated.

FMEA Working Sheet
Component/Item Name:
Function

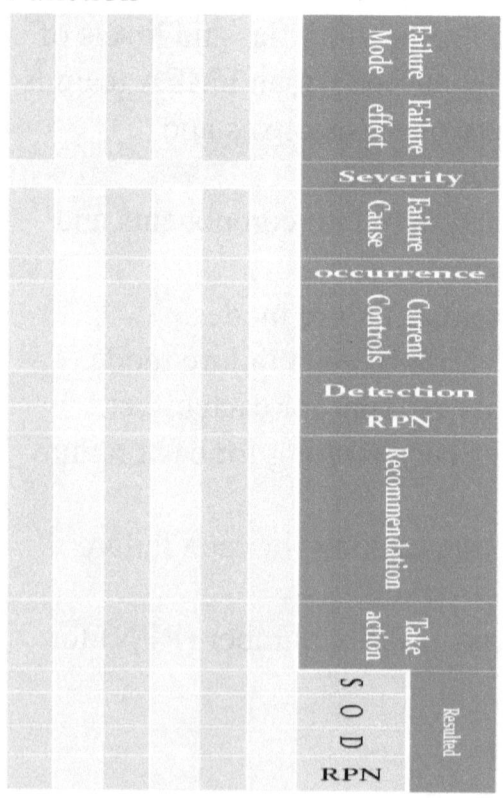

Step.1 Review the Process or Product
If the team is considering a product, they should review the engineering drawing of the product.
If the team considering a process, they should review the operation flowchart.
This is to ensure that everyone has the same understanding about the process or product.
For a product, they should physically see the product and operate it.
For a process, they should physically walk through the process exactly as the process flows.

Step.2 Breakdown the System into Components and Sub-components
If the system is a large system, like a water system that supplies an industrial process, the pump can be a critical component inside the system. A motor pump is a critical subcomponent because its failure can break down the entire process. The motor pump should be broken down into more subcomponents that are likely to fail and will affect the system, such as the motor's bearings and the rotor shaft. The FMEA will be used to prevent the probability of failure for each component or subcomponent.

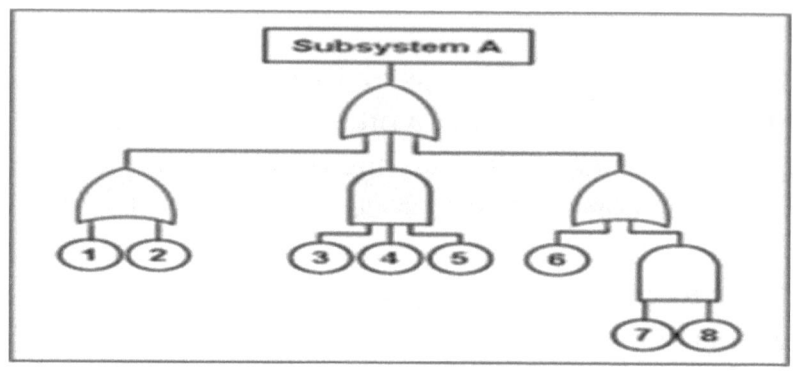

Step.3 Brain Storm Potential Failure Modes

When everybody in the group has a comprehension about the item or the cycle, colleagues should start contemplating the potential failure modes that could influence the process or the item quality. Zeroing in ought to be on the various components (individuals, material, management, strategy… and so on). When the conceptualizing is finished, the thoughts ought to be sorted out by gathering them into like classes. There are numerous approaches to assemble failure modes, they can be gathered by kind of failure (electrical, mechanical, client made). Where on the item or cycle the failure happens.

Main Rules of Brainstorm:

Try not to remark on, judge or study thoughts at the time they are advertised. Empower inventive and odd thoughts. The objective is to wind up with an enormous number of thoughts; and assess thoughts later. Every thought ought to be recorded and numbered precisely as offered, on a flip outline.

Expect to generate at least 50 to 60 concepts in a 30-minute brainstorming session.

Failure Mode & Effect Analysis FMEA

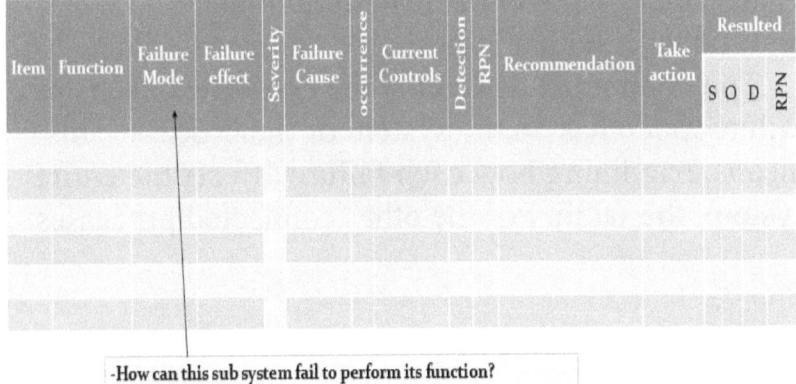

- How can this sub system fail to perform its function?
- The Way the failure occurred
- What will the operator see?

Step.4 List Potential Effects for Each Failure Mode

For a portion of the failure modes, there might be one impact, while for different modes, there might be a few impacts. This data must be through that it will take care of into the task of the risk ranking for every one of the failures.

Tips:

One failure mode could have several effects. For example, an electrical cutoff in the home could stop the refrigerator and damage food or prevent you from doing work on the computer.

Several failure modes could have one effect. A dead car battery or tire failure has the same effect on your vehicle – it will be difficult to make it to work on time with such a failure early in the morning.

The team must determine the end-effect each failure mode has on the system or the process. This means examining how each failure affects the entire system, the facility or the other connected processes.

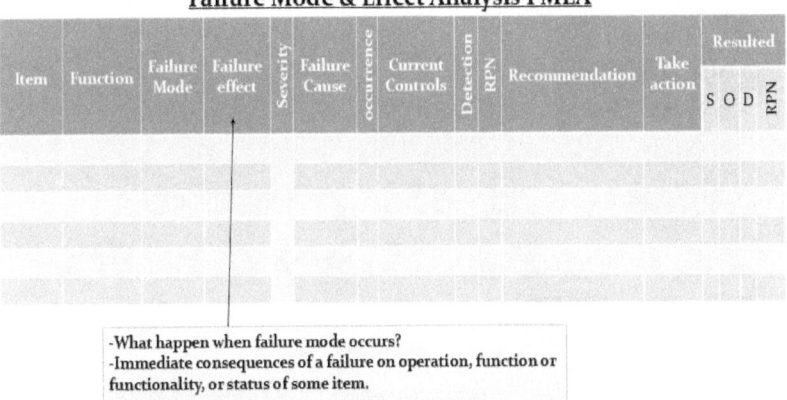

Steps 5-7 Assign Severity, Occurrence, and Detection Rankings

Each of these three rankings is based on 10-point scale, with 1 being the lowest ranking, and 10 the highest.

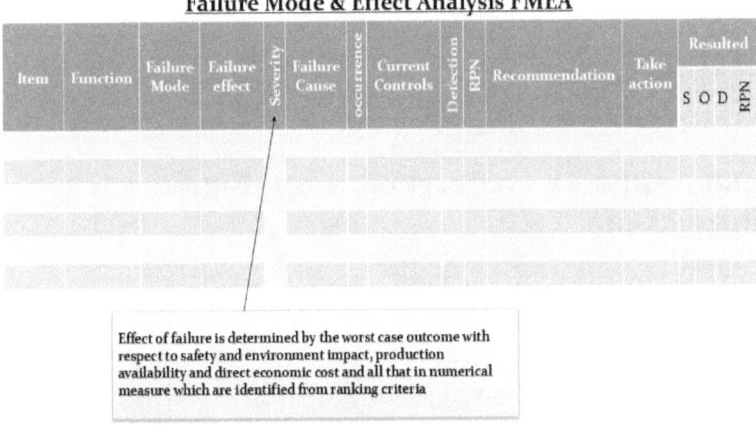

Failure Mode & Effect Analysis FMEA

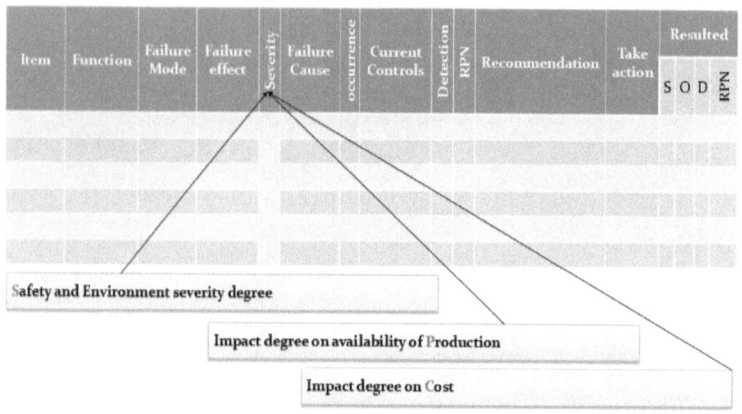

Severity Ranking Criteria

Description of Failure Effect	Effect	Ranking
No reason to expect failure to have any effect on Safety, Health, Environment or Mission.	None	1
Minor disruption of production. Repair of failure can be accomplished during trouble call.	Very Low	2
Minor disruption of production. Repair of failure may be longer than trouble call but does not delay Mission.	Low	3
Moderate disruption of production. Some portion too of the production process may be delayed.	Low to Moderate	4
Moderate disruption of production. The production process will be delayed.	Moderate	5
Moderate disruption of production. Some portion of production function is lost. Moderate delay in to High restoring function.	Moderate to High	6
High disruption of production. Some portion of production function is lost. Significant delay in restoring function.	High	7
High disruption of production. All of production function is lost. Significant delay in restoring High function.	Very High	8
Potential Safety, Health or Environmental issue. Failure will occur with warning.	Hazard	9
Potential Safety, Health or Environmental issue. Failure will occur without warning.	Hazard	10

Step.6 Assign an Occurrence Ranking for each Failure Mode

The best technique for deciding the occurrence ranking is to utilize real information from the process. This might be as failure history. At the point when real failure information is not accessible, the group must gauge how frequently a failure mode may happen, the group can improve gauge on how likely a failure mode is to happen and at what recurrence by knowing the expected reason for failure. When the potential causes have been distinguished for the entirety of the failure modes, an occurrence ranking can be appointed regardless of whether the failure information are not existed.

Failure Mode & Effect Analysis FMEA

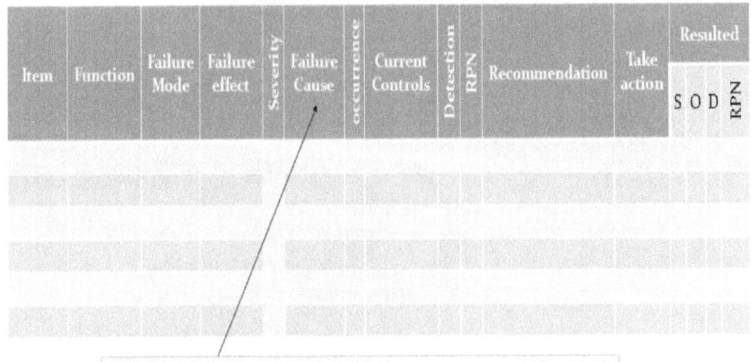

For each failure mode there may be several failure causes. Assign a Cause for each failure mode.
Select only potential failure to get failure causes.
Use Why Why Technique to get the root causes.
Identifying the failure cause can be the second option to determine the occurrence if no data is available in the form of failure logs.

Failure Mode & Effect Analysis FMEA

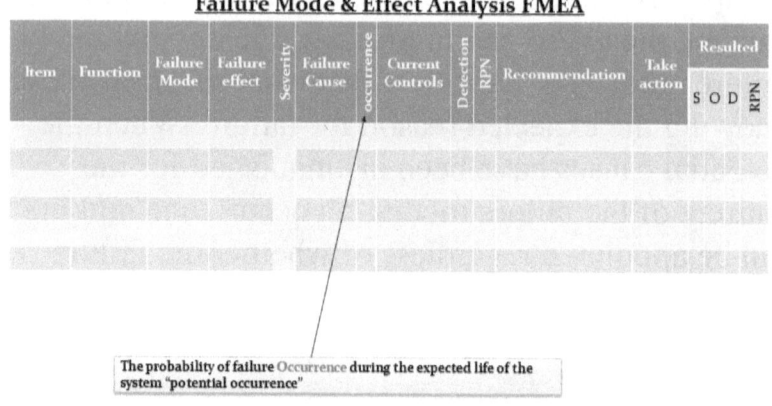

The probability of failure Occurrence during the expected life of the system "potential occurrence"

Occurrence Ranking Criteria

Rank	Freq	Description
1	1/10,000	Remote probability of occurrence; unreasonable to expect failure to occur
2	1/5,000	Low failure rate; similar to past design that has, in the past, had low failure rates for given volume or load
3	1/2,000	Low failure rate; similar to past design that has, in the past, had low failure rates for given volume or load
4	1/1000	Occasional failure rate; similar to past design that has, in the past, had similar failure rates for given volume or load
5	1/500	Moderate failure rate; similar to past design that has, in the past, had moderate failure rates for given volume or load
6	1/200	Moderate failure rate; similar to past design that has, in the past, had moderate failure rates for given volume or load
7	1/100	High failure rate; similar to past design that has, in the past, had high failure rates that have caused problems
8	1/50	High failure rate; similar to past design that has, in the past, had high failure rates that have caused problems
9	1/20	Very High failure rate; almost certain to cause Problems
10	1/10	Very High failure rate; almost certain to cause Problems

Operating hours based on the automotive industry benchmark.
Ranking can be determined based on historical data or similar system benchmarking

Step.7 Assign a Detection Ranking for each Failure Mode and/or Effect

To start with, the current control ought to be recorded for the entirety of the failure modes, or impacts, and afterward the recognition rankings appointed. In the event that one failure mode or impact has a few causes, recognition and occurrence rankings ought to be relegated dependent on these causes. At the point when potential causes are disposed of, the danger of failure is brought down.

In case the application is for an equipment maintenance, current control methods can be the current preventive maintenance program and or the current detection methods (condition monitoring program).

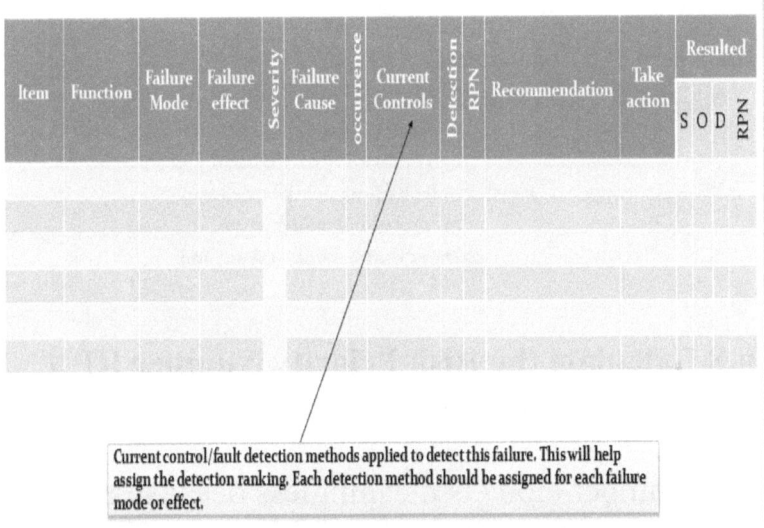

Current control/fault detection methods applied to detect this failure. This will help assign the detection ranking. Each detection method should be assigned for each failure mode or effect.

Failure Mode & Effect Analysis FMEA

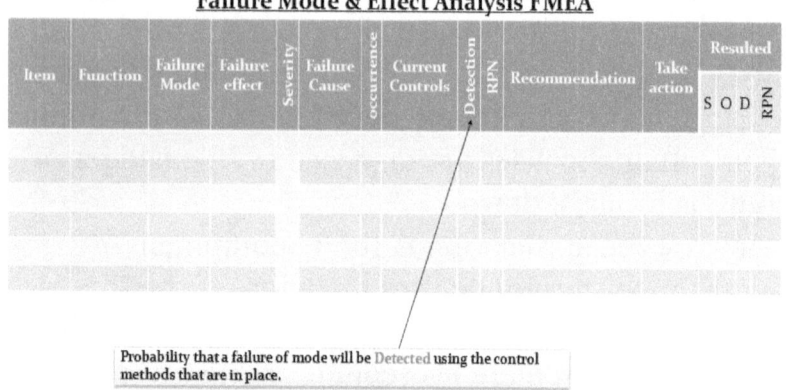

Probability that a failure of mode will be Detected using the control methods that are in place.

Detection Ranking Criteria

Rank	Description
1-2	Very high probability of detection
3-4	High probability of detection
5-7	Moderate probability of detection
8-9	Low probability of detection
10	Very low probability of detection

Step.8 Calculate the Risk Priority Number RPN

Risk Priority number= Severity x Occurrence x Detection.

This number alone is meaningless because each FMEA has a different number of failure modes and effects. However, it can serve as a gauge to compare the revised RPN once the recommended actions has been instituted.

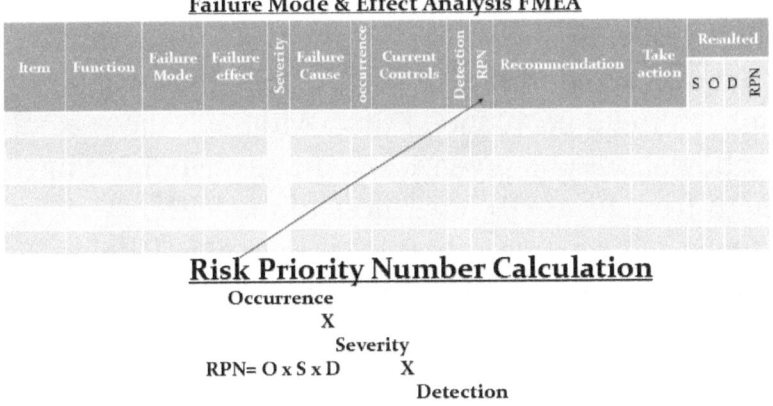

Risk Priority Number Calculation

$$RPN = O \times S \times D$$

Occurrence X Severity X Detection

What is RPN?

The Risk Priority Number (RPN) methodology is a technique for analyzing the risk associated with potential problems identified during a Failure Mode and Effects Analysis (FMEA).

RPN Calculation Benefits:

- Contribute in Risk Assessment.
- Compare components to determine priority for corrective action. Components with higher RPN are given more attention.
- Assessing the risk priority number.

Each potential failure mode or effect is rated in each of these three factors on a scale ranging from 1 to 10. By multiplying the ranking a risk priority number RPN can be determined for each potential failure mode and effect.

The RPN will range from 1 to 1000 for each failure mode. It is used to rank the need for corrective action. Those failure modes with the highest RPN number should be attended first. Although the special attention should be given when the severity ranking is high from (9 to 10) regardless of the RPN.

Once a corrective action is takes, a new RPN is determined. This new RPN is called the resulting RPN.

Step.9 Prioritize the Failure Modes for Action

Failure modes ought to be organized by positioning them all together, from the most elevated danger need number to the least. Odds are that you will find that the standard 80/20 principle applied with the RPNs. A Pareto chart should be created.

The group should now choose which thing to work for. Typically, it assists with setting a cutoff RPN (cutoff point), where any failure modes with a RPN over that point are taken care of. Those beneath the cutoff are disregarded until further notice.

Tips:

High-risk numbers should be given attention first; then you can pay attention to the severity rankings. Thus, if several failure modes have the same risk priority number, that failure mode with the highest severity should be given more priority. If severity number is the same, those failures with higher occurrence should be given more priority and so on.

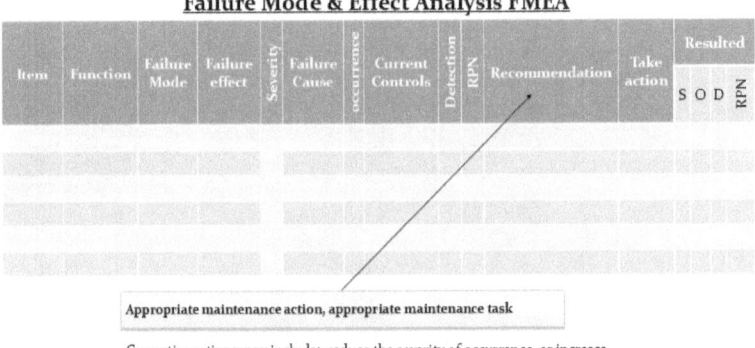

Corrective actions may include: reduce the severity of occurrence, or increase the detection probability

Step.10 Take Actions to Eliminate or Reduce the High-Risk Failure Modes

This is organized using the problems-solving approaches and implement actions to reduce or eliminate the high-risk failure modes.

Often the easiest way to make an improvement to the product or process is to increase the detectability of the failure, thus lowering the detection rate.

Increase the detection rate can be done though assigning a schedule PM action, use a proper condition monitoring program or consider a mistake proofing method in the design. For example, ac computer software will automatically warn in case of low disk space.

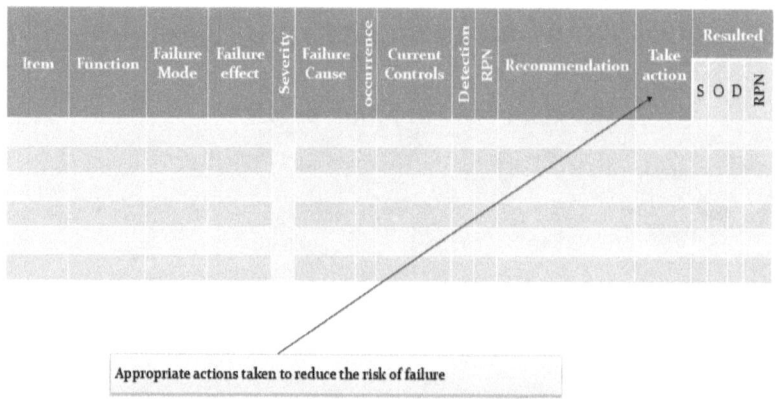

Appropriate actions taken to reduce the risk of failure

Step.11 Calculate the Risk Priority Number RPN as the High Risk is Removed

When moves have been made to lessen the danger need number, another positioning for the seriousness, event, and discovery ought to be determined. What's more, a subsequent RPN is determined. Desire is in any event 50 rate decrease in RPN with the FMEA approach.

There will consistently be a potential for failure modes to happen. The inquiry the organization must pose is how much relative danger the group is eager to take. That answer may depend on the business and the reality of the disappointment. For instance, in the atomic business, there is a little edge for mistakes, they can't hazard a calamity happening. In different enterprises, it might be worthy to face the high challenge.

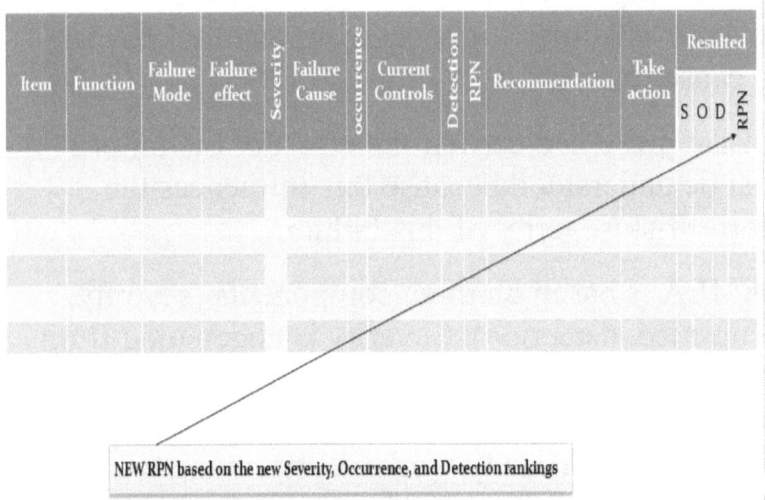

NEW RPN based on the new Severity, Occurrence, and Detection rankings

FAILURE MODES, EFFECTS, AND CAUSES

For each failure mode, causes have to be determined, and the final root cause must be identified because from there you are going to work on improving/minimizing the failure. Preventing failure from occurring require knowing the cause to tackle it.

Like presented earlier, the effect of the failure must be understood because this influences the priority/risk/severity of this failure.

FMEA is based on three components (severity, occurrence, detection). Severity is understood from how the failure will affect the system. Occurrence is determined from the causes, and detection is the factor that u can easily work on improving to improve the Risk Priority Number and reduce risks. Most of FMEA projects I have worked on rely on improving detection. Reducing severity require re designing the system to minimize the effect of failure, which is very difficult and high cost. So, the most easy and affordable part is to improve detection to reduce occurrence thus minimize the risk of failure.

It's very important to understand that running FMEA during the design process is very important to reduce severity. Also, redundancy reduce the overall risk of failure. Eg. A standby generator that will work to supply the system with electricity when the transformer is down. A standby pump that will operate immediately when the main pump fails...etc.

Japanese people don't like the idea of redundancy. They think it's a costly option in the production process. Also takes space and waste money. They believe if everything is having redundancy or standby, you will never think about improving the system and the process. Also, for high risk/safety equipment like elevators, a failure could result in death or injury. Same in Aircraft, nuclear and aerospace industries. The type of assets being used in those high-risk industries require a continuous system of assessing risks and improving the process. FMEA provide a robust process of assessing risks and minimizing potential failures.

Here are example of some failure mechanisms/symptoms for several machines/equipment:

Electric Transformer

Failure Mode	Failure Mode	Failure Cause
Voltage regulation issue	Tap changers	Mechanical wear
High noise/ Corona	Cable insulation	Overload
High oil temp	Fins	Blockage
Oil leakage	Tank gasket damage	Wear

Vehicle

Failure Mode	Failure Mode	Failure Cause
Car won't start المحرك لا يدور	Fuel delivery pump طلمبه البنزين	Pump motor wear/fail عطل فى الطلمبه
	Ignition coil الموبينا	Burn coil احتراق الموبينا
	Blockage in fuel filter السداد فى فلتر البنزين	Dirties in the tank شوائب فى التنك
	Cam belt سير الكاتينه	Broken القطاع
	Control unit وحده التحكم	Failed/Water got in عطل نتيجه تسرب مياه
Poor idling	Actuator sensor (IAVC)	Life time
	ECU Control unit	Failed
	Injectors الرشاشات	Blockage due to dirties in fuel السداد نتيجه شوائب فى البنزين
	Spark plugs البوجيهات	Wear تآكل أو تلف
	Fuel filter فلتر البنزين	Dirties causing blockage

Failure Mode	Failure Mode	Failure Cause
Car performance (acceleration problem)	Spark plugs	Wear
	Air Filter	Dirty
	Fuel filter	Blocked
	Fuel pump	Weak
	Speed sensor	Faulty
	ECU	Faulty
	Compression ratio	Worn cylinder head
		Cylinder head gasket
		Wear piston rings
		Internal wear

Generator

Failure Mode	Failure Cause
Engine is running, but no AC output is available	1. One of the circuit breakers is open. 2. Fault in generator. 3. Poor connection or defective cord set. 4. Connected device is bad.
Engine runs good at no-load but "bogs down" when loads are connected.	1. Short circuit in a connected load. 2. Engine speed is too slow. 3. Generator is overloaded. 4. Shorted generator circuit. 5. Clogged or dirty fuel filter.

Failure Mode	Failure Cause
Engine will not start; or starts and runs rough.	1. Start switch in off (O) position. 2. Fuel valve is in "Off" position. 3. Failed battery. 4. Low oil level. 5. Dirty air cleaner. 6. Clogged or dirty fuel filter. 7. Fill fuel tank. 8. Drain fuel tank and carburetor; fill with fresh fuel. 9. Spark plug wire not connected to spark plug. 10. Bad spark plug. 11. Water in fuel. 12. Flooded.

Failure Mode	Failure Cause
Engine is running, but no AC output is available	1. One of the circuit breakers is open. 2. Fault in generator. 3. Poor connection or defective cord set. 4. Connected device is bad.
Engine runs good at no-load but "bogs down" when loads are connected.	1. Short circuit in a connected load. 2. Engine speed is too slow. 3. Generator is overloaded. 4. Shorted generator circuit. 5. Clogged or dirty fuel filter.

Belt Conveyor

Failure Mode	Failure Cause
All Portions Of The Conveyor Belt Run To One Side At A Given Point On The Structure.	1. One Or More Idlers Immediately Preceding Trouble Not At Right Angles To The Direction Of Belt Travel. 2. Conveyor Frame Or Structure Crooked. One Or More Idler Stands Not Centered Under Belt. 3. Sticking Idlers. 4. Buildup Of Material On Idlers. 5. Structure Not Level And Belt Tends To Shift To Low Side.

Failure Mode	Failure Cause
A Particular Section Of The Conveyor Belt Runs To One Side At All Points On The Conveyor.	1. Belt Not Joined Squarely. 2. Bowed Belt 3. Worn Edge
Severe Wear On The Pulley Side Of The Conveyor Belt.	1. Slippage On Drive Pulley 2. Material Spills Between Belt And Pulley. 3. Material Build Up At Loading Point Until Belt Is Dragging 4. Sticking Idlers 5. Excessive Tilt To Trough in Idlers 6. Bolt Heads Protruding Above Lagging 7. Bottom Cover Too Thin.

Failure Mode	Cause
The Conveyor Belt Runs To One Side For Long Distance Along The Bed.	Load Being Placed On Belt Off Center. Conveyor Frame Or Structure Crooked.
Fasteners Pull Out	1. Wrong Type Of Fastener Or Fasteners Not Tight 2. Tension To High 3. Heat 4. Tandem Drive Poorly Compensated 5. Inadequate Convex Curve Radius
Drive Pulley Spins	1. Belt Tension. 2. Drive Pulley lag
Conveyor belt breaks	1. Belt Overloaded 2. Splice Failure 3. Belt Wear From Age
Belt Moves Sideways, After Making A Complete Cycle	Belt Was Not Squared When Spliced.

Failure Mode	Cause
Excessive material carry back on the belt's top cover causes build up on the snub pulley and return idlers	No belt scraper Bad belt scraper Bad lagging
Material splicing separated	Material buildup migrate and grind into the top cover and into small imperfections in a belt splice

Failure modes at Spectrum Vibration analyzers

Failure Mode: Frequency in terms of RPM	Most likely causes	Other possible causes and remarks
1x RPM	Unbalance	1) Eccentric journals, gears or pulleys 2) Misalignment or bent shaft - if high axial vibration 3) Resonance 4) Reciprocating forces 5) Electrical problems
2x RPM	Mechanical Looseness	1) Misalignment if high axial vibration 2) Reciprocating forces 3) Resonance 4) Bad belts if 2x RPM of belt
3x RPM	Misalignment	Usually a combination of misalignment and excessive axial clearances (looseness)

Less than 1x RPM	Oil whirl (less than ½ RPM)	1) Bad drive belts 2) Background vibration 3) Sub-harmonic resonance
Synchronous (A.C. Line Frequency)	Electrical Problems	Common electrical problems include broken rotor bars, eccentric rotor, unbalanced phases in poly-phase systems, unequal air gap.
2x Synch. Frequency	Torque pulses	Rare as a problem unless resonance is excited
Many times RPM (harmonically related freq.)	Bad gears Aerodynamic forces Hydraulic forces Mechanical looseness Reciprocating forces	Gear teeth times RPM of bad gear Number of fan blades times RPM Number of impeller vanes times RPM May occur at 2,3,4 and sometimes higher harmonics if severe looseness
High frequency (not harmonically related)	Bad anti-friction bearings	1) Bearing vibration may be unsteady-amplitude and frequency 2) Cavitations, recirculation and flow turbulence cause random, high frequency vibration. 3) Improper lubrication of journal bearings (friction excited vibration) 4) Rubbing

Failure Modes at Ultrasound Detector

Ultrasound Noise (Symptom) – Failure Mode	Failure Cause
Steady regular buzzing or frying sound when measuring high-medium voltage devices	Corona
A buzzing and intermittent crackling sound when measuring high-medium voltage devices	Arcing
A violent sound of an electrical arc with an abrupt start and stop when measuring high-medium voltage devices	Tracking
High frequency spectrum when measuring the bearing	Bad bearing lubrication
Friction sound when measuring compressors reciprocating valves	Valve wear

Example Failure Mode, Effects, and Causes:

Ex.1 Centrifugal Fan

Failure mode	Failure Effect	Failure Effect (System)	Failure Effect (End)	Failure cause Level 1	Root cause
Fan operate with high vibration level	Equipment damage/breakdown	Unexpected plant shutdown	Major production losses	Bearing fails	Poor Maint
	Equipment damage/breakdown	Unexpected plant shutdown	Major production losses	Housing wear	Poor Maint
	Equipment damage/breakdown	Unexpected plant shutdown	Major production losses	Unbalance fan blade	Poor Maint
	Equipment damage/breakdown	Unexpected plant shutdown	Major production losses	Looseness in foundation	Poor Maint
	Equipment damage/breakdown	Unexpected plant shutdown	Major production losses	Shaft wear	Poor Maint

Ex.2 Transformer

Item name	Failure mode	Failure Effect (local)	Failure Effect (System)	Failure cause	Failure Cause	Root cause
Oil	1.Short circuit in transformer	Functional stop	Production losses	Particles in the oil	Overheated	Bad Maintenance
		Functional stop	Production losses	Water in the oil	Overheated	Bad Maintenance
					Aging	
Tap Changes	2-Can't change voltage level	Functional stop	Production losses	Mechanical damage	Wear	Life time/ maintenance

Ex.3 Water System

Function	Functional failure/failure modes	Causes
Provide water to the industrial process	Total loss of pressure, volume & flow	Pump failed Motor failed Valve out of position

Electric Motor

Function	Functional failure/failure modes	Causes
Drive the water pump	Burn out	Circuit Breaker tripped Bearing seized Insulation Rotor Insulation Stator

Motor Bearing

Failure mode	Failure Cause	Sources of failure/causes	Causes
Bearing seized, this include bearing, seals, lubrication	Lubrication	Contamination	Supply dirty Sealing failed
		Wrong type	Procedure wrong Supply information wrong
		Tool little	Human error Procedure error
		Too much	Human error Procedure error

Final Table

Failure effect			Severity			Causes	Root Cause	Occurrence	Current fault detection methods	Detection	RPN	Actions
Local	sys	end	S	A	C	Seal failed	Seal failed					
Motor shutdown	System shutdown	TPL				Procedure wrong	Lack of training					
						Human error						
						Human error						

Risk=Probability x Severity

Probability or frequency	Consequence or Severity		
	(1) Low	(2) Medium	(3) High
(1) Low	1 L	2 L	3 M
(2) Medium	2 L	4 M	6 H
(3) High	3 M	6 H	9 H

It's important to design your own matrix

FMEA added the Detection which is something you can easily control and manage to reduce risks.

A CASE OF RELIABLE IMPROVEMENT

IMPROVING THE RELIABILITY OF A PUMP STATION SYSTEM

After analyzing the system, the team came to the most common failures are coming from an electric motor exists in this pump station.

Breaking down the components of the electrical motor

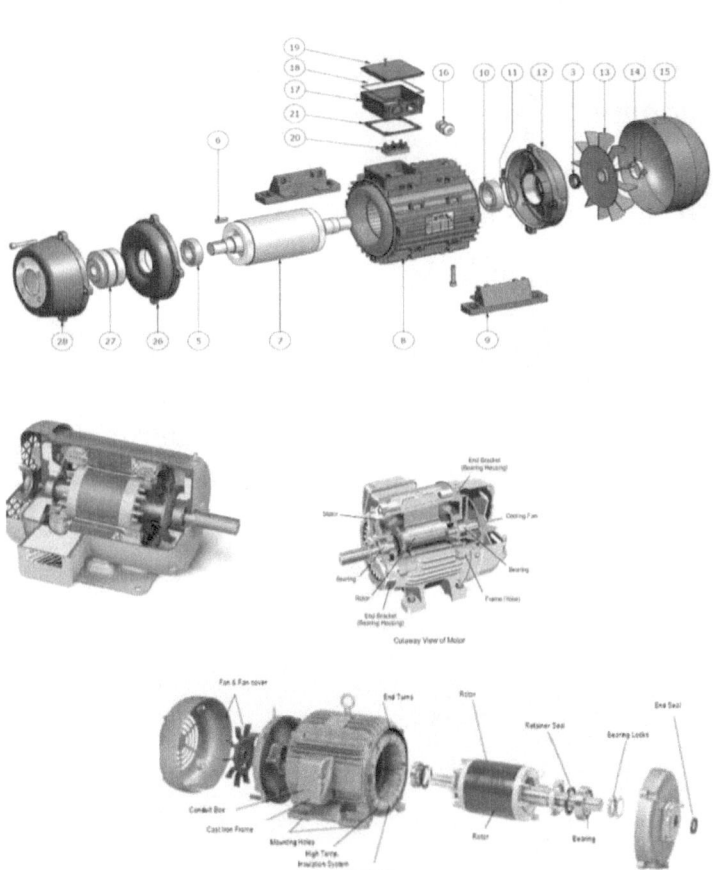

Equipment Information

Equipment Type   : DC Electric Motor

Technical Specs   : 50KW

Function   : Supply electricity to the water pump

System   : Water pump station, supply water for the industrial process

Availability of standby system: No

The electric motor is considered critical because a failure causes high production losses. The pump supplies the industrial process with the water required for the crystal manufacturing. Using a standby was not possible at the moment because it will require a change in the design which is very costly.

Electric Motor Failure Components

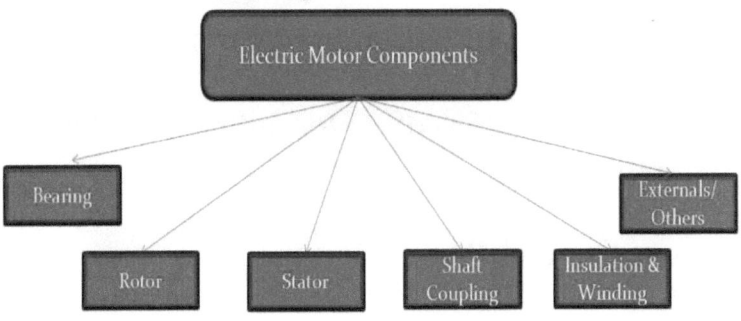

Electric motor sub-components

Electric Motor Failures

- ✓ You can group failures by mechanical, electrical…etc.
- ✓ More than 40% of motor failures are bearing failures.
- ✓ More than 25% of motor failures are stator failures.

Current Control/Prevention methods

PM Activity	Component/Item	PM Level
Check air gap between the rotor and stator with feeler gages	Rotor	Annually
	Stator	Annually
Inspect for misalignment	Coupling	Annually
Inspect for excessive wear. Check for proper type, hardness, conductivity, and fit in brush holders. Check holder spring pressure with a small scale. In most instances, pressure should be 2 to 2 1/2 lbs per sq in. of brush cross-sectional area	Brushes and commutators	Annually
Visual Inspection	Motor mounts, bolts, nuts, and tightening	Annually
Clean / Blow out dirties with compressed air. Remove any dust, chemicals, grease or dirt	Motor body and the cooling fan	Annually

Failure Log History

Working Condition= 24 hrs.

Failure type	Causes	Frequency (5 yrs)
Overheating & overloading	Bearing damage	6
Poor output	Motor inefficiency due to stator problems	2
High vibration	misalignment	2

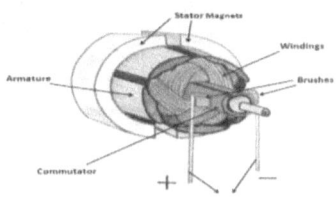

Failure Analysis

Component Name & Function: Bearing, reduce friction of the rotating shaft

Failure Mode	Failure Effect	Severity	Failure Causes	Failure Cause	Failure Causes	Occurrence	Current control methods	Detection	RPN
High vibration. Bearing Failed/ Seized	overload & overheat	8	Improper lubrication or grease	In correct type of lubricant		1	NA	8	64
				Wrong procedure	Lack of training	1		10	80
				In sufficient lubricant	Lack of maintenance	5		8	320
	High repair cost due to possible shaft damage		Improper mounting	Lack of tools		1	NA	1	8
				Lack of training	No standard	2		10	160
			Parts damage	Aging	Lack of maintenance		Annual inspection of shaft/coupling misalignment		
			Shaft misalignment	Lack of maintenance		2		8	128

Recommendation	Take actions	Result		
		S	O	D	RPN
1. Monitor bearing health condition with vibration and thermograph 2. Use standard procedures for bearing mounting and lubrication	Monitor vibration monthly, use standard work process and train workers them. Keep auditing the maintenance procedures	8	1	2	16
		8	1	2	16

Risk Assessment Using FMEA: A Case of Reliable improvement | Mohammed Soliman

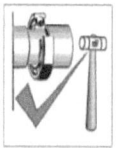
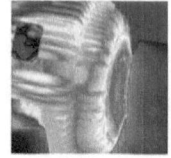
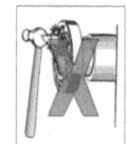
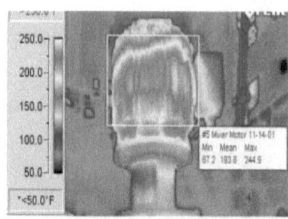

Bearing damage due to incorrect fitting

Component Name & Function: Coupling & shaft, transmit the movement

Failure Mode	Failure Effect	Severity	Failure cause	Failure Cause	Failure Cause	Failure Cause	Occurrence	Current controls	Detection	RPM
Misalignment	Equipment shutdown to avoid bearing damage and expensive repairs	8	Physical damage	Wear	Aging	Bad maintenance	1	Visual inspection	6	48
			Improper manufact	Bad type/material			1		10	80
			Improper installation	Bad Maint			1	Misalignment check annually	8	64
			Corrosion	Bad Maint			1		10	80
Recommendation			Take actions				Result			
							S	O	D	RPN

Flexible Coupling

Self Aligning
Ball Bearing

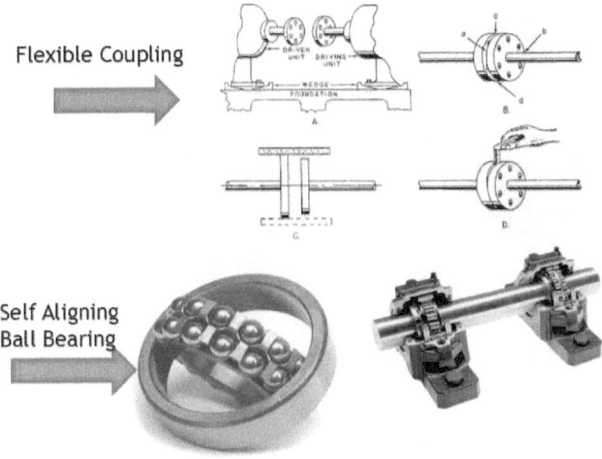

Precision Maintenance "Shaft Alignment"

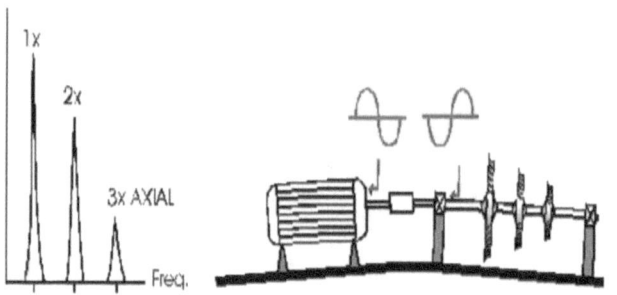

Angular misalignment detection using vibration analysis

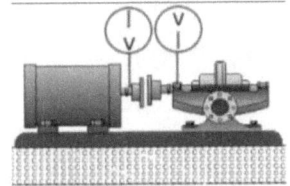

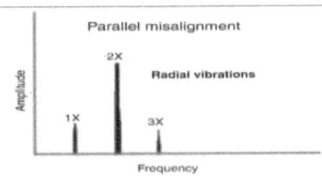

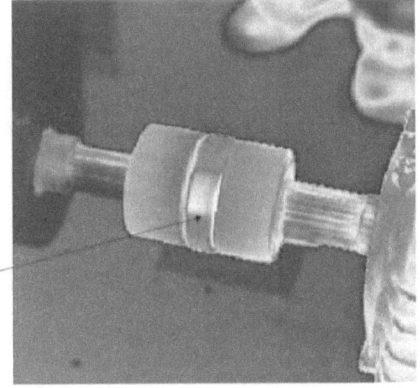

Coupling problem

Name & Function: Stator, generate electricity, carry current, retain armature

Failure Mode	Failure Effect	Severity	Failure cause	Failure Cause	Failure Cause	Occurrence	Current Controls	Detection	RPN
Stator defect	Motor inefficiency, High cost to repair	8	Eccentricity	High temp	Wear/Aging/Lack of maintenance	5	Check air gap between the rotor and stator with feeler gages	10	400
				Corrosion					
				High temp					
				Contamination					
			Short lamination	High temp					
				Corrosion					
				High temp				10	400
				Contamination					
			Loose iron	High temp					
				Corrosion					
				High temp					
				Contamination					

Recommendation	Take actions	Result			
		S	O	D	RPN
Monitor motor stator condition & temp with a proper method	Use vibration analysis & infrared thermograph analysis	8	1	1	8
		8	1	1	8

Stator defects

Eccentricity
Short lamination
Loose iron

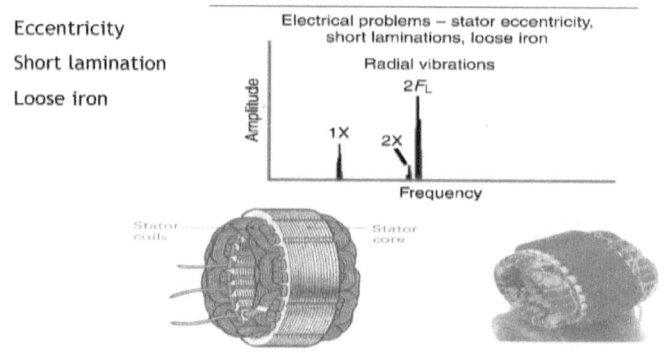

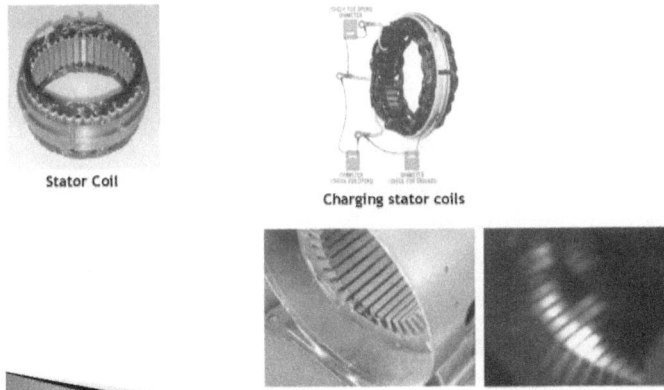

Stator Coil

Charging stator coils

Minor core damage

Core damage as it appears in infrared

A D.C. motor consists of a rectangular coil made of insulated copper wire wound on a soft iron core. This coil wound on the soft iron core forms the armature. The coil is mounted on an axle and is placed between the cylindrical concave poles of a magnet.

Component Name & function: Rotor, is the moving part of the motor

Failure Mode	Failure Effect	Severity	Failure cause	Failure Cause	Failure Case	Cause	Occurrence	Current Controls	Detection	RPN
Rotor defect	Bearing damage, motor re build, high repair cost	8	Eccentric rotor	Imbalance	Wear	Aging	1	Checking air gap between the rotor and stator with feeler gages	10	80
				Thermal stress						
				Assembly problem	Bad maintenance					
				Soft foot or poor base						
			Broken rotor bars	Imbalance	Wear	Aging	1		10	80
				Thermal stress						
				Assembly problem	Bad maintenance					
				Soft foot or poor base						

Recommendation	Take actions	Result			
		S	O	D	RPN
Monitor rotor condition with a proper method	Use vibration analysis to detect rotor defects	4	1	2	8
		4	1	2	8

The rotor is a moving component of an electromagnetic system in the electric motor, electric generator, or alternator. Its rotation is due to the interaction between the windings and magnetic fields which produces a torque about the rotor's axis. When the coil rotates, the shaft attached to it also rotates and thus it is able to do mechanical work.

Damaged motor shaft

Detection of Different Electric Problems

The following are some terms that will be required to understand vibrations due to electrical problems:

F_L = electrical line frequency (50/60 Hz)

F_s = slip frequency = $\dfrac{2 \times F_L}{P}$ – rpm

F_p = pole pass frequency = $F_s \times P$

P = number of poles.

Rotor problems

1. Broken rotor bars
2. Open or shorted rotor wind
3. Bowed rotor
4. Eccentric rotor

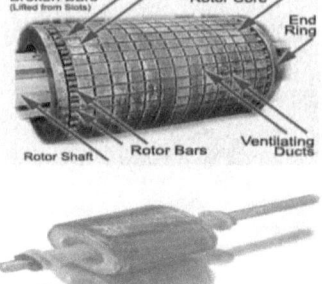

Defect rotor bars

A. Rotor Defects.
Broken rotor bars
Eccentric rotor

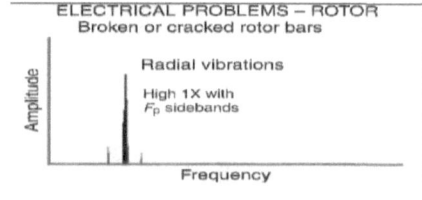
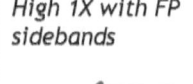

High 1X with FP sideband

Broken rotor bars

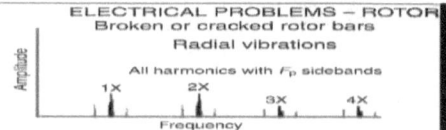

Broken rotor bars
All harmonics with FP sidebands

Example:

Motor Speed Synchronous = 1800RPM
Motor Speed Actual = 1770RPM
No of poles=4
FL=60hz
FP= 2xFL/P-RPM*P=2hz

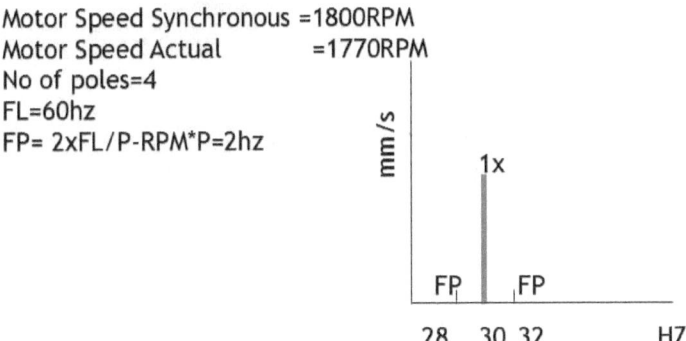

B. Eccentric rotor

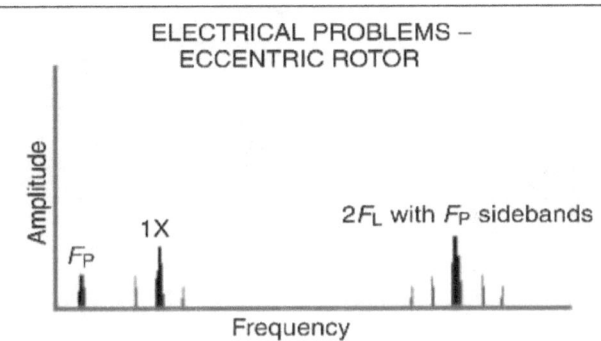

Component Name & Function: Insulation & winding, carry current

Failure Mode	Failure Effect	Severity	Failure cause	Failure Cause	Failure Cause	Occurrence	Current Controls	Detection	RPN
Winding failure/shortage	Motor failure	8	Overheat	Bad maint	Wear	2	Basic measurements include voltage	6	96
			Moisture	Bad maint					
			Contamination						
			Insulation breakdown						
			High vibration	Bearing	Bad maint	3		10	240
				Misalignment		2		10	160
			Voltage surges	aging	Bad maint	1		10	80

Recommendation	Take actions	Result			
		S	O	D	RPN
Monitor machine vibration on regular basis	Use vibration analysis to monitor bearing condition and shaft misalignment	8	3	1	24
		8	2	1	16

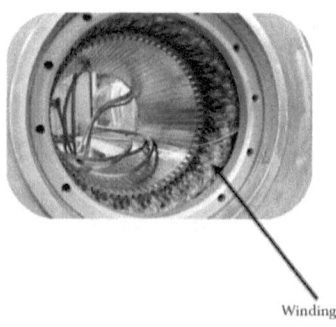

Winding

Insulation

Component Name & Function: Fan, keep the motor temp down in order for the motor components to perform well

Failure Mode	Failure Effect	Severity	Failure cause	Sources of failure	Failure Cause	Occurrence	Current Controls	Detection	RPN
Fan failure	Overheating and lead to expensive repair	8	Corrosion	Environmental issue	Aging	1	Visual inspection & cleaning	2	16
			Physical damage	Crash		1		2	16
				Carless handling					
			Foreign material build up	Lack of cleaning	Bad maintenance	1		2	16
						1		2	16

RPN Analysis for Electric Motor Failure Components

Part/Item	RPN	Part/Item	RPN
Bearing	64	Rotor	80
	80		80
	320	Winding & Insulation	96
	8		240
	160		160
	128		80
Shaft & Coupling	48	Cooling Fan	16
	80		16
	64		16
	80		16
Stator	400	**Total**	**2624**
	400		

A cutoff point of RPN 160 can be set because this will achieve over 50% improvement to risk number.

A cutoff point of RPN 160 can be set because this will achieve over 50% improvement to risk number.

Expected Total Risk Priority Number after applying the corrective actions:

RPN Reduction % = $R_{initial} - R_{revised} / R_{initial}$
  = 2624-1040/2624
  = 60%

Vibration Analysis

Increase inspection reduce the risk of failure

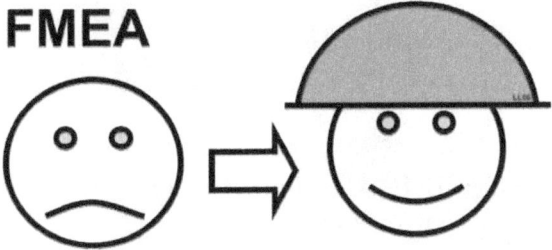

The improvements that yielded success included using vibration analysis to detect electrical issues and using infrared analysis to detect mechanical damages.

FMEA AND CONTINUOUS IMPROVEMENT

Each step is a FMEA toward the target, so please repeat the FMEA wheel!

An FMEA process can trigger a number of such actions to improve a product's service or maintenance processes. They include, but are not limited to:

- ✓ Increase the detection rate of high-risk failures using a proper technique to monitor conditions.
- ✓ Increase the inspection rate for a specific component or part.
- ✓ Modify the routine maintenance program.
- ✓ Increase the frequency of replacing a specific spare part.
- ✓ Modify the preventive maintenance schedule.
- ✓ Change a spare part supplier.
- ✓ Redesign a specific part in the system – or redesign the whole system.
- ✓ Use different types of materials or spare parts.

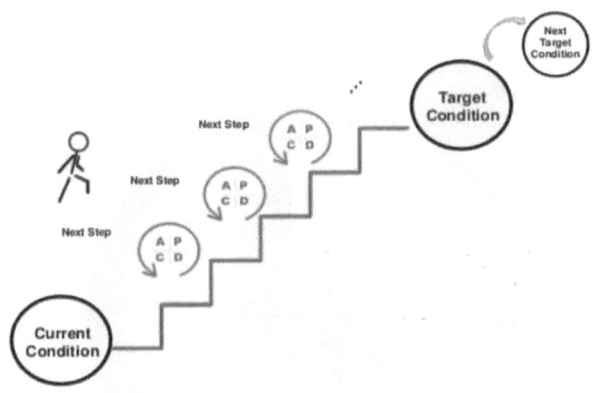

Does FMEA Sound Like a Standalone Tool?

Failure mode and effects analysis work with the other Quality tools to maximize a product's reliability. It doesn't work as a standalone tool. For example, to determine occurrence ratings, FMEAs rely on the failure log history, and the documentation process also is important. Problem-solving techniques like "five whys," brainstorming, fault-tree analysis and Pareto analysis must be engaged. These techniques will help determine potential failure modes; assign the severity, occurrence and detection rankings; and provide solutions or actions to eliminate those failures.

How to get the why?

= Go and see =

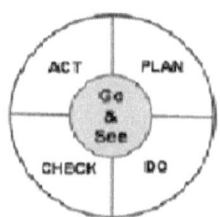

- ✓ Remember to brainstorm potential failure modes.

- ✓ Use Pareto to prioritize actions, and which failure mode to tackle first. It doesn't make sense to make improvements to all failures at once! This will increase project cost and results may not be seen to managers as cost effective.

- ✓ Use 5Whys to solve problems and find the real root cause of issues.

Remember Plan-Do-Check-Act cycle is an endless cycle. You will never be perfect until repeating it again and again.

APPENDIX.1 BRAINSTORMING

Because FMEA is not a standalone tool and must be integrated with some other tools, the two most important tools mentioned usually with any FMEA process is Brainstorming and Root Cause Analysis.

You need brainstorming to brainstorm potential failure modes for each component of the product. And you need the Root Cause Analysis to find the cases of the failure.

What is Brainstorming?
A quality tool for issues comprehending that ought to be a contributor to any issues tackling practice.

Brainstorming is a conceptualizing technique that is a notable strategy for creating countless thoughts in a brief timeframe period. It fills in as an instrument for recognizing issues and causes.

To energize thoughts, no thought ought to be scrutinized or remarked when advertised. Every thought ought to be recorded and numbered, precisely as offered on a flip outline.

Hope to produce in any event 50 to 60 thoughts in a 30 minutes meeting to generate new ideas.

> Brainstorming helps to collect the data needed for any Total Quality Management Process

Brainstorming Rules

- Try not to remark on, judge or evaluate thoughts as advertised.
- Support inventive and odd thoughts.
- An enormous number of thoughts is the objective.
- Assess thoughts later.

> When the brainstorming session is over, the ideas should be reviewed, similar ideas combined, and ideas that do not seem to fit eliminated.

Brainstorming is a group problem-solving method. It taps people creative ability to identify and solve problems, and brings out a lot of ideas in a very short time. Because it is a group process, it helps builds people as human beings. For example, brainstorming encourages individual members to contribute to the group and to develop trust for the other members.

What is required for Brainstorming?
1. A gathering ready to cooperate
You may feel it is unthinkable that the gathering you work with will never be a group. Be that as it may, conceptualizing can be a key to manufacture a group! Besides, it is an incredible apparatus for the gathering which is as of now cooperating.

Who ought to be remembered for the gathering?
Each and every individual who is worried for the issue for two reasons: the thoughts for every individual who worried about the difficult will be accessible for the talk. Second, those individuals can take a functioning part in tackling the issue. In that manner they can be got the chance to help the arrangement.

2. A pioneer
- The principal parts of the pioneer are:
- Give some direction so conceptualize will deliver thoughts
- Authority over the gathering to keep them on target.
- Empower individuals' thoughts and interest.
- Set the individual objectives aside to help the gathering.

3. A gathering place

A spot where there is no interference or interruption. In certain plants, bunches utilize a foreman office, a region on the creation floor, or even a meeting room.

4. Resources or Tools
Flipcharts, markers, and white sheets

How Does it Work?
- ✓ Pick a subject for the talk.
- ✓ Ensure that everybody comprehend what the issue or the subject is.
- ✓ Every individual is to take a turn an express one thought. On the off chance that somebody can't consider anything, the individual in question says "pass". In the event that somebody thinks about a thought when it isn't his turn, he may wright it down on a paper and use it at his next turn.
- ✓ Record every thought precisely as communicated.
- ✓ Try to compose all thoughts and don't dismiss any.
- ✓ Energize wild thoughts, they may trigger another person's reasoning.
- ✓ Hold criticism until after the session.
- ✓ The main goal is quantity and creativity.
- ✓ A little laughter is fun and healthy but don't overdo. It is O.K to laugh with someone but not at them.
- ✓ Allow few hours or days for further thoughts (if needed). The first brainstorm on a subject will stimulate people to start thinking, but an incubation period allow mind to release more creative ideas and thoughts.

Example: Conducting a Brainstorming Session
Conceptualizing the Causes of a Defective Capacitor

This gathering incorporates five individuals: Samy, the pioneer; Farouk; Mohammed; Gamal, the recorder, and Ahmed. Since they have been meeting for just a brief timeframe and the individuals have not had a lot of involvement in conceptualizing, the pioneer needs to do a large portion of crafted by keeping them on target. As the gathering picks up experience, different individuals should start to share crafted by the initiative.

Samy: I believe it's an ideal opportunity to conceptualize for reasons for flawed capacitors. Gamal, since you are acceptable at flip outline, would you be able to help us there?

Gamal: Yes obviously.

Samy: Let us set a 15 minutes time cap for the meeting. What's more, remember the standards: We will go around from individual to another, one thought at time. Try not to stress if your thought sound unusual. All things considered, regardless of whether your thought is a wild one, it might invigorate another person. No assessments. We will have a lot of time thereafter to take a gander at the thoughts. Alright would you say you are prepared? (Everyone concurs.)

Farouk, your turn.

Farouk: Vendor (Gamal records, VENDOR).

Mohammed: I have seen gouges in some of them. Furthermore, I feel that a mark outwardly implies something breaks or gets pressed or somehow wrecked inside.

Samy: Mohammed, you are stating "gouges". Is that right?

Mohammed: No, I mean gouges show us there is an issue inside.

Samy: Can we abridge it to peruse: "Imprints show inside issue"?

(Mohammed gestures "O. K"

Samy: Gamal, it's your turn.

Gamal: I figure I will relax.

Ahmed: The prompts the capacitor now and then doesn't get fastened well. So, makes it resemble a blemished capacitor.

Gamal: How would I compose that? "Binding of leads"?

Ahmed: Yup that is O.K.

Samy: My turn. I will expand on Farouk's concept of "seller". May be its just one of them that is actually the issue and not every one of them. Gamal state "One Vendor"

Farouk: Seems to me the state of AX12's is the issue. They help me to remember the latrine seats spread. (Much Laughter).

Samy: Let's return to the subject. Farouk, may have something there. So Samy express "State of AX12's".

Tips and Techniques for Completing the Session: Nudging Techniques

Also called "prodding' technique. Sooner or later the downpours of ideas in the brainstorm dries up. What do you do to get it going again? Or what do you do with the silent member who doesn't participate?

Empowering Thoughts: making preparations once more "priming the pump again"

If the brainstorming session seems to slow down, the leader may suggest piggybacking. Piggybacking is building on others' ideas. For example, if one of the team members has suggested the vendor as a cause of the problem, another one might say "one vendor" not all of them could be the reason of the problem.

Another technique is to suggest opposites. For example, too much & too little.

Dealing with the silent member

When a member of the group doesn't speak up, the best way to deal with this is to be patient.

Sometimes a person will be quite for a meeting after meeting then he will open up. It will be then very exciting, so give this person a time. Maybe he/she will be quiet, but will serve the group with some other ways.

A simple effective method to bring the silent member, is to remind the whole group that when each person's turn comes in the brainstorm, he or she just says "Pass" if not ready with an idea. That gets people of the hock but it also breaks the sound barrier. They hear their own voices and participate by saying "Pass."

The direct question is another method, but you must use it with care.

Something like" Mohammed, you know the process well, do you have a suggestion or input here?"

The Second Session

After the initial brainstorm and sometime for further thinking, it's a good idea to have another session to capture more ideas. These ideas come into mind as the group member think about the problem and consider what was said in the first session.

Two different ways to deal with the subsequent meeting:

Assemble all gathering and give them a period cutoff of 10-12 minutes for extra thoughts. Similar principles applied as in the primary meeting.

Post the conceptualizing sheets in the territory of the working environment with the goal that it will permit individuals who work in a similar region to contribute regardless of whether they are not a normal individual from the critical thinking gathering. In that manner they believe they are not forgotten about.

Finishing a talk

How would you ensure that conceptualizing has secured all potential reasons for an issue?

Here and there the arrangement lies in a pursuit lab, where just a high prepared master gets an opportunity of uncovering it. Frequently, in this way, the arrangements are directly close to home.

Regardless of whether you don't tackle the issue immediately, you can ensure that you have secured all the overall zones of potential causes. Make a rundown of the overall regions, and ensure that you're gathering or group has analyzed all of them.

Such a list would include a number of subjects. There are some major factors that go into an operation: management, machine, method, material, environment, and people.

Machine include: the type of the machine, the maintenance, and the setting.

Materials are the elements that come to the process, whether they are raw material, sub-assemblies, components, or partially processed materials.

Method concerns the process itself.

Environment is important too, humidity, dust, and other climate problems that may affect the process.

Management concerning procedures standards, training and leadership.

Finally, the Person doing the job. Factors connected with the person could be training, eyesight, and level of skills.

Other general areas may also apply to the problem. Such as money, process, and other errors.

Troubles and How to Manage Them
You are stepping on my turf!
It will be hard for one of the gathering individuals to be associated that he is the reasons with this issue. For instance, the plan engineer is going to the meeting, and the reason for the issue came out to be in the plan cycle.

Train your group, and create them. It is important to clarify that we are not here to accuse anybody. What's more, we are at times dazzle with our issues so we need others to look on it. We will in general observe just a contributor to the issue that is the reason the causes might be covered up. It involves forthcoming.

Analysis
Fabricate a positive climate in the gathering. Censure issues not individuals. Ensure that thoughts not people are assessed. Ensure that errors are not announced and never show up in anybody individual life.

The troublesome part
A few individuals are hard to manage in the gathering. They blabber, they get off track, they scrutinize individuals not thoughts or they destroy thoughts. How would you manage him?

Be firm however cordial. Converse with him secretly and clarify how his way is diverting the gathering work. Give the troublesome part an uncommon employment to accomplish for the gathering. Try not to battle him. At the point when he gets the gathering off course, re back the discussion to the ordinary theme tenderly.

Generally troublesome individuals turned into the most grounded help of the gathering, or they leave.

APPENDIX.2: CAUSE AND EFFECT DIAGRAM

5 Whys

5 whys is the method of completing the cause and effect diagram to tackle the root cause of the problem and prevent recurrent failure. It was originated in Japan. Japanese people believe that by asking 5 whys you can figure out the root cause of the problem and find the solution. However, it doesn't have to be 5 it can be 7 or 8.

Toyota

Toyota does not have a six-sigma program. Six sigma is based on complex statistical quality analysis tools. It is a surprise for people to realize how Toyota has achieved this level of quality without the use of six sigma for quality.

Most of problems don't call for complex statistical analysis, but instead require detailed problem solving. This requires a level of detailed thinking and analysis that is all too absent from most companies in day-to-day activities.

Level of Problem	Countermeasure
There is an oil on the shop floor	Clean up the oil
Because the machine is leaking (Why?)	Fix the machine
Because the gasket has deteriorated (Why?)	Replace the gasket
Because we bought gaskets made of inferior material (Why?)	Change gasket specifications
Because we got a good deal/price on those gaskets (Why?)	Change purchasing policy
Because the purchasing gets evaluated on short-term cost saving (Why?)	Change the evaluation policy for purchasing agent

5 whys is a method to pursue the deeper, systematic causes of a problem to find correspondingly deeper countermeasures

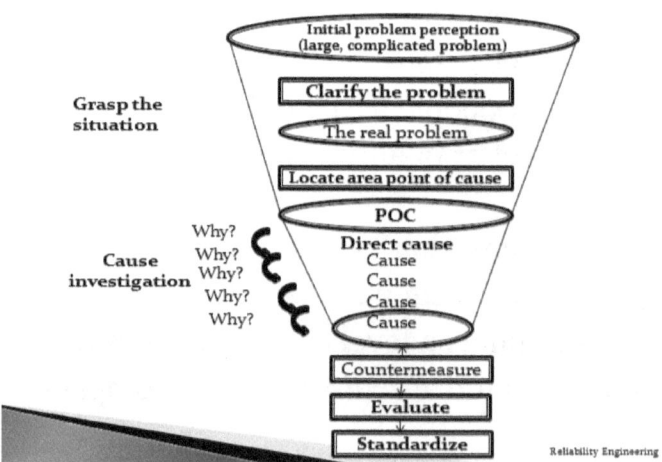

Reliability Engineering

Example:

Q1: Why did the customer not buy the product?
A: The salesperson did not persuade him to buy.
Q2: Why did the salesperson not persuade the customer to buy?
A: The salesperson was not good enough.
Q3: Why was the salesperson not good enough?
A: The sales person has not been trained in sales.
Q4: Why has the salesperson not been trained in sales?
A: It was not considered necessary.
Q5: Why was training not considered necessary?
A: Sales are only a small part of the job.

Drawing Cause and Effect Diagram (Fishbone Diagram)

Using a fishbone diagram while brainstorming possible causes helps
you to focus on the various possibilities. Some useful categories:

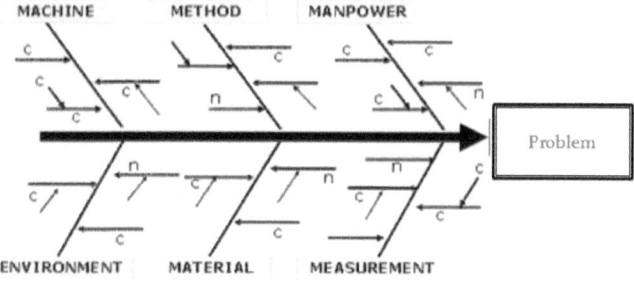

The main problem is entered in the nose. The bones originally had only "4Ms". Once all problems were reduced to one of the four: man, machine, material, or method. Eventually, measurement was added to highlight how critical it is to have an understanding of the reliability and accuracy of the measuring system. Environment was added to make people consider the location of an equipment and the impact of its surroundings on the operation. Design and instruction can also be a good reason to add. Fishbone diagram take inputs from brainstorming sessions. Those possible countermeasures are the ideas that people give during the brainstorming sessions. Dig deep into the details at the Gemba (the place where real work happens) is necessary to absorb the real situation, perform diagnosis, make analysis, talk to people that are involved directly in the work and base the solution on facts.

REFERENCES

KTH Electrical Engineering.

Liker, J. K. (2003). Toyota way: 14 Management Principles. New York: MacGraw-hill.

Raymond J. Mikulak, Robin McDermott. (2008). The Basics of FMEA. Productivity Press; 2 editions.

Soliman, M.H.A. 2014. Analyzing Failure to Prevent Problems. Industrial Management.

Soliman, M.H.A. 2020. Industrial Applications of Infrared Thermography. KDP and Lulu Press.

Soliman, M.H.A. 2020. Ultrasound Analysis for Condition Monitoring. KDP.

Soliman, M.H.A. 2020. Practical Guide to FMEA: A Proactive Approach to Failure Analysis.

Soliman, M.H.A. 2020. Vibration Basics and Machine Reliability Simplified: A Practical Guide to Vibration Analysis.

Soliman, M.H.A. 2020. Machine Reliability and Condition Monitoring.

Soliman, M.H.A. 2020. Brainstorming for Problems Solving: How Leaders Can Achieve a Successful Brainstorming Session.

Robert T. Amsden and Davida M. Amsdenand. (1998). SPC Simpliefied: Practical steps to quality. Productivity Press; 2 editions.

ABOUT THE AUTHOR

Mohammed Hamed Ahmed Soliman is an industrial engineer, consultant, university lecturer, operational excellence leader, and author. He works as a lecturer at the American University in Cairo and as a consultant for several international industrial organizations.

Soliman earned a bachelor of science in Engineering and a master's degree in Quality Management. He earned post-graduate degrees in Industrial Engineering and Engineering Management. He holds numerous certificates in management, industry, quality, and cost engineering.

For most of his career, Soliman worked as a regular employee for various industrial sectors. This included crystal-glass making, fertilizers, and chemicals. He did this while educating people about the culture of continuous improvement.

Soliman has lectured at Princess Noura University and trained the maintenance team in Vale Oman Pelletizing Company. He has been lecturing at The American University in Cairo for 6 year and has designed and delivered 40 leadership and technical skills enhancement training modules.

Soliman is a member at the Institute of Industrial and Systems Engineers and a member with the Society for Engineering and Management Systems. He has published several articles in peer reviewed academic journals and magazines. His writings on lean manufacturing, leadership, productivity, and business appear in Industrial Engineers, Lean Thinking, and Industrial Management. Soliman's blog is www.personal-lean.org.

Also, by Mohammed Hamed Ahmed Soliman

https://www.amazon.com/~/e/B00NEY7BRE

Recommended reads: